ILLUSTRATED NOTES ON BIOMOLECULES

ILLUSTRATED NOTES ON BIOMOLECULES

MOHAMMAD FAHAD ULLAH

PARTRIDGE

To order additional copies of this book, contact
Toll Free 800 101 2657 (Singapore)
Toll Free 1 800 81 7340 (Malaysia)
orders.singapore@partridgepublishing.com

www.partridgepublishing.com/singapore

Contents

Preface

When we analyze the elemental composition of our non-living surroundings such as the Earth crust, oceans or atmosphere and our (human) tissues or for that matter any living organism including plants, animals and organisms of microbial world, it is observed that all living organisms are made of same elements as are the non-living entities. So the perplexing thought arises as to what grants the living organism, the "life". The response to such a thought comes from the ability of these organisms to convert the simple elements into an array of complex molecules which could be easily differentiated from the non-living forms, and are the fundamental basis of life. These complex molecules that give life to living forms are called as "Biomolecule". All these organic molecules which are the structural and functional basis of living forms are the product of biological activities.

This book is the first edition of the title "ILLUSTRATED NOTES ON BIOMOLECULES" and presents the basic knowledge on biomolecules. The book has a student centric approach of learning process with the presentation of the knowledge in an easy to understand language in a manner that should induce the intent of self-learning among the students. An attempt has been made to break up the complex wall of knowledge into facile bricks and let the students join the bricks themselves. An easy understanding of the subject evokes interest towards advance learning.

Mohammad Fahad Ullah, Ph.D
Prince Fahd Research Chair
Department of Medical Laboratory Technology
FAMS, University of Tabuk, KSA

This book is dedicated to my beloved father who will ever remain an inspiring soul for all my deeds!

Foreword

I am immensely excited after reading "Illustrated Notes on Biomolecules", a book on the very 'basis' of life by a young and very enthusiastic scholar, Dr. Mohammad Fahad Ullah. It is indeed fascinating to realize that the same structural blocks, that make up a non-living structure, stack together to form complex macromolecules  that define 'life'. These macromolecules, or the 'biomolecules', are very clearly organized in nature, and Dr. Fahad Ullah has done a commendable job of organizing the information in this book to clearly explain the hierarchy in the organization of such biomolecules. The topics in the book appear at a pace that should be very convenient to the students just getting exposed to this subject material. The basic information is explained in sufficient detail before a simple break down of more complex structures.

This should be an excellent source of information for students in high school and those pursuing under-graduate studies. A clear understanding of fundamental physiological biomolecules is the key to success for all those who are aiming for a successful career in health and medical sciences. I remain very optimistic about the success and popularity of this book.

Aamir Ahmad, PhD
Mitchell Cancer Institute
University of South Alabama
Mobile, AL, USA

Learning objectives

Rationale & outcome

All activities are assessed for their merit by measuring the end outcomes and these are marked as successful only when the outcome is productive. Similarly, learning process is also guided by learning objectives which measure the outcomes with regard to knowledge and understanding that has been obtained during the course of the learning process. The following are the learning objectives which the reader of this book will be able to gauge at the end:

- Understand the characteristics of biomolecules with regard to the fundamental concepts of functional groups and chemical bondings;
- Identify and define different types of biomolecules with emphasis on macromolecules;
- Describe the important structural features of biomolecules;
- Classify carbohydrates in terms of structural complexities & functional significance;
- Explain the basis of amino acid classifications and the structural organization & functions of proteins;
- Appreciate the structural & functional diversities in lipid molecules;
- Explain the difference between DNA and RNA and variations in their composition;
- Acknowledge vitamins as essential small biomolecules and the associated deficiency disorders.
- Understand the significance of metabolites in biochemical pathways

Biomolecules

Introduction

Biochemistry is a branch of science which deals with thousands of chemical reactions occurring in a living cell. Since these reactions are essentially responsible for the characteristics of the cells that make it a living entity, they are termed as biochemical reactions. The molecules of the cell which participate in these biochemical reactions so as to form the structural and functional basis of life are known as *biomolecules*. The four elements which are most abundant in living organisms accounting for the 99% of the mass of most cells include carbon (C), hydrogen (H), oxygen (O) and nitrogen (N). The bulk of biomolecules are made up of these four elements though other elements such as phosphorus and sulphur along with trace elements are also constituents of some biomolecules.

The master element around which the chemistry of biomolecules revolves is carbon which has a unique ability to form C to C single, double or triple bonds to give stable structures of linear chains, branched chains and cyclic organization. Carbon can also form single bonds with the other three major elements, such as single bonds with hydrogen atoms and single/double bonds with oxygen and nitrogen atoms, forming complex structures of varied biomolecules.

Classification

Biomolecules present in a cell are numerous, however only the important ones have been included in the current description.

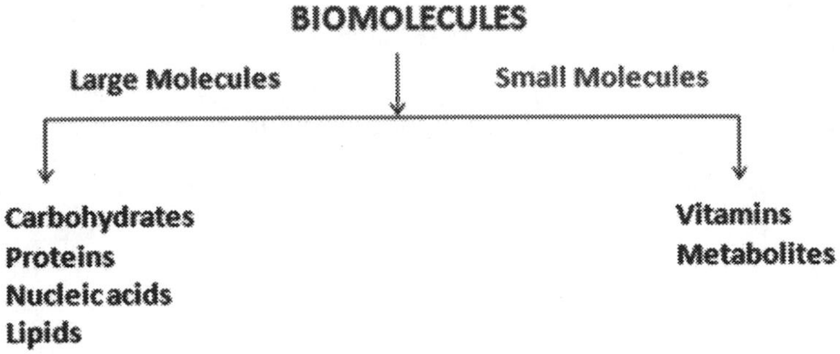

BIOMOLECULES

Large Molecules   Small Molecules

Carbohydrates   Vitamins
Proteins   Metabolites
Nucleic acids
Lipids

There are four major classes of biomolecules which are referred as macromolecules and these include:

- Carbohydrates
- Proteins
- Lipids
- Nucleic acids

The other classes of biomolecules include small molecules such as:

- Vitamins
- Metabolic elements (metabolites).

Carbohydrates

Carbohydrates are the basic fuel molecules of the cell. It may break them down to yield chemical energy or use them as raw materials to produce other molecules; e.g. glucose.

Proteins

Proteins are large molecules (polymers) made up of smaller subunits called amino acids. The proteins are of vital interest within the cell since these: (i) participate in the building of cell structure; e.g. actin/ myosin (muscle protein) and (ii) are responsible for the cellular functions such as biocatalysts (enzymes) to speed up a biochemical reaction.

Lipids

Lipids are fatty substances that play a variety of roles in the cell. Some are stored for use as high-energy fuel; e.g. triacylglycerides and others like phospholipids serve as essential components of cell structure such as plasma membrane.

Nucleic acids

Nucleic acids are polymer of smaller subunits called as nucleotides. Polymerized nucleotides form two kinds of nucleic acids which are DNA (**D**eoxyribo**N**ucleic **A**cid) and RNA (**R**ibo**N**ucleic **A**cid). Nucleic acid such as DNA serves as genetic material whereas RNA molecules are engaged in the processing of genetic information.

Vitamins

Vitamins are non-caloric organic micronutrients that are essential for growth and metabolism due to their participation in a myriad of biochemical reactions. These are required in small quantities and are classified as water soluble vitamins (B complex and C) and fat soluble vitamins (A, D, E and K).

Metabolites

The sum of all the chemical transformations taking place in a cell occurs through a series of enzyme-catalyzed reactions called as metabolic pathways. The constituents of a metabolic pathway from starting compound (precursor) to the final end product and metabolic intermediates are called as metabolites; e.g. glucose, fructose 6-phosphate and pyruvate (are metabolites of glycolysis).

Learner's Notes

Functional Groups

Biomolecules such as carbohydrates, proteins, lipids, nucleic acids and others are built upon a very stable hydrocarbon skeleton. The hydrogen atoms within these skeletons may be replaced with specific groups of atoms which result in forming a variety of organic compounds and are principally responsible for the chemical characteristics of the resulting compounds. Such specific groups of atoms are known as functional groups. These may include hydroxyl groups (as in alcohols or carbohydrates), amino groups (as in amines or amino acids), carbonyl groups (as in aldehydes and ketones), carboxyl groups (as in amino acids and fatty acids), thiol group (as in amino acid cysteine) and phosphate groups (as in nucleic acids). Some biomolecules are polyfunctional as they contain two or more varied kinds of functional groups; e.g. carbohydrates have hydroxyl groups as well as a carbonyl group; amino acid cysteine which has an amino group, a carboxyl group as well as a thiol group.

Carbohydrates

OH
|
—C—
|
Alcohol

O
‖
—C—H
Aldehyde

O
‖
—C—
Ketone

**Amino acids
(Proteins)**

NH₂
|
—C—
|
Amine

O
‖
—C—OH
Carboxyl

— SH
Thiol

**Fatty acids
(Lipids)**

O
‖
—C—OH
Carboxyl

Nucleic acids

O
‖
—O—P—O⁻
|
O⁻
Phosphate

Covalent bonds in large biomolecules (macromolecules)

Considering the four principal biomolecules, three among them, namely proteins, nucleic acids and some carbohydrates (polysaccharides) are polymeric molecules which are made up of smaller subunits (monomers). Lipids though not a polymer is also considered a large molecule. These lipid molecules include triacylglycerides or sphingolipids in which fatty acids are covalently esterified with alcohols such as glycerol or sphingosine.

The monomeric subunits in these large molecules are linked together by means of specific covalent bonds which are formed as a product of condensation reaction through dehydration. In polysaccharides, the monosaccharide subunits are linked by *glycosidic* bonds. Amino acids in proteins are linked by *peptide* bonds whereas nucleotides in nucleic acids are bonded with *phosphodiester* linkages. In lipids such as triacylglycerides, fatty acids are covalently bonded to glycerol by *ester* linkages.

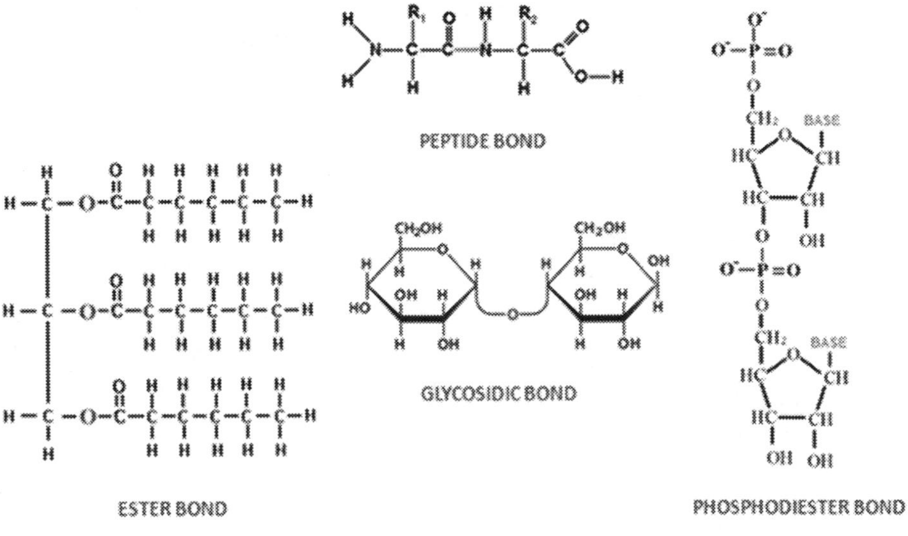

PEPTIDE BOND

GLYCOSIDIC BOND

ESTER BOND

PHOSPHODIESTER BOND

Learner's Notes

Carbohydrates

Carbohydrates form a very large group of naturally occurring organic compounds which are the most abundant biomolecules on Earth. In general, the carbohydrates may be defined as polyhydroxy aldehydes or ketones or substances that give such molecules on hydrolysis. Most of the carbohydrates have sweet taste and therefore these are termed as saccharide, a word derived from Greek sakcharon which means sugar. Carbohydrates are classified as simple sugars (monosaccharides or disaccharides) and complex sugars (oligosaccharides or polysaccharides) with their common names ending with a suffix "-ose".

Classification of carbohydrates

Monosaccharides

The monosaccharides are the simplest carbohydrates which are represented by the general formula $(CH_2O)_n$, where n=3 or more. Depending on the number of carbon atoms, they are classified as trioses (3C), tetroses (4C), pentoses (5C), hexoses (6C) and heptoses (7C). The chemical structure of a monosaccharide has a polyhydroxy carbon chain with a single carbonyl group. The monosaccharides with carbonyl group as an aldehyde (-CHO) are called as aldoses whereas those with carbonyl group as a ketone (>C=O) are called as ketoses.

Aldose

Ketose

D–Glucose   D–Fructose

An aldose sugar with the least number of carbon atoms as 3 is referred as an aldotriose (glycerladehyde) and a ketose sugar with same carbon number as a ketotriose (dihydroxyacetone). Similarly, other monosaccharides of aldose and ketose series are aldotetrose and ketotetrose; aldopentose and ketopentose; aldohexose and ketohexose as described below:

Number of carbon atoms	Aldoses		Ketoses	
Three	Aldotriose	Glyceraldehyde	Ketotriose	Dihydroxyacetone
Four	Aldotetrose	Erythrose, Threose	Ketotetrose	Erythrulose
Five	Aldopentose	Ribose, Arabinose, Xylose, Lyxose	Ketopentose	Ribulose, Xylulose
Six	Aldohexose	Glucose, Mannose, Galactose	Ketohexose	Fructose

Stereoisomers

In stereochemistry, a carbon is said to be asymmetric (chiral) if it is bonded to four different groups of atoms. Most monosaccharides except dihydroxyacetone contain one or more chiral carbon atoms and thus occur as optical isoforms. The simplest monosaccharide glyceraldehyde, has a single asymmetric carbon atom (the central carbon), so two stereoisomers (optical isomers) exists, denoted as D- and L-glyceradehyde, which are mirror images of each other. Other monosaccharides derive D- and L-configuration from the configuration of asymmetric carbon of glyceraldehyde isomers corresponding to the configuration of their asymmetric carbon which is furthest from the carbonyl carbon (aldehyde or ketone group). If the configuration of the asymmetric carbon furthest from the carbonyl carbon is same as D-glycerladehyde, the monosaccharide is designated as D isomer. However, if the configuration of the asymmetric carbon furthest from the carbonyl carbon is same as L-glycerladehyde, the monosaccharide is designated as L isomer.

In case a monosaccharide has n number of asymmetric carbon, the number of stereoisomers possible for this monosaccharide is given as 2^n. For example, glucose which is an aldohexose has 4 asymmetric carbons and so the number of stereoisomers possible for glucose is $2^4 = 16$. Eight of these are D-isomers and eight are L-isomers. Similarly, fructose, a ketohexose has 3 asymmetric carbons, so the number of stereoisomers for fructose is $2^3 = 8$. The D- and L- stereoisomers are mirror images of each other and are called as **enantiomers.** Thus, D and L glucose are enantiomers.

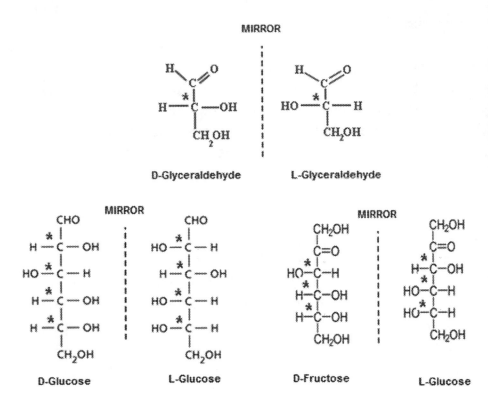

* Asymmetric carbon atoms

Epimers are stereoisomers of monosaccharides which differ in the orientation of hydroxyl group (-OH) at a single asymmetric /chiral carbon. For example; D-Glucose and D-Galactose are epimers as they differ in the orientation of hydroxyl group (-OH) at a single asymmetric carbon (C4). Similarly, D-Glucose and D-Mannose are epimers as they differ in the orientation of hydroxyl group (-OH) at a single asymmetric carbon (C2).

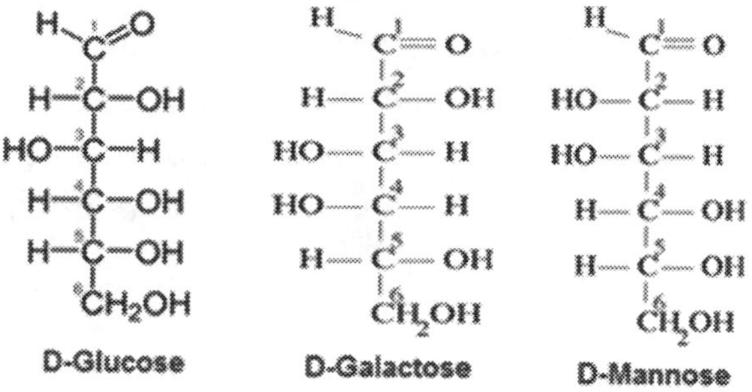

D-Glucose   **D-Galactose**   **D-Mannose**

In general, the structures of monosaccharides are represented in an open chain form (as shown above). Such open chain structures are called as Fischer formula of the sugars. However, aldehyde or ketone group can react with a hydroxyl group within the chain to form a covalent bond. The reaction between an aldehyde and the hydroxyl group creates a hemiacetal whereas the one between a ketone and a hydroxyl group form a hemiketal. Such reactions result in the cyclization of the open chain to from ring structures such as pyranose ring (6 membered) or furanose ring (5 membered). These ring structures are referred to as Haworth projection formula of the sugars.

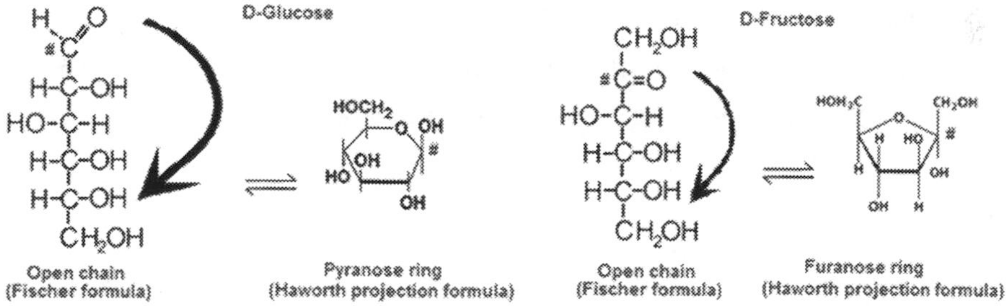

Open chain
(Fischer formula)

Pyranose ring
(Haworth projection formula)

Open chain
(Fischer formula)

Furanose ring
(Haworth projection formula)

During the conversion from straight-chain form to the cyclic form, the carbon atom containing the carbonyl oxygen, called the anomeric carbon, becomes a stereogenic center with two possible configurations: The resulting possible pair of stereoisomers is called

anomers. In the *α-anomer, the −OH* substituent on the anomeric carbon lie below the plane of the ring. The alternative form in which the *−OH* group lies above the plane of the ring is called the *β-anomer. e.g. α, D- Glucose and β, D-Glucose are anomers as they differ in the orientation of the hydroxyl group at anomeric carbon (C1) of the ring structure.*

Monosaccharides are reducing agents

Sugars that contain a free aldehyde or ketone group can reduce ferric (Fe^{3+}) or cupric (Cu^{2+}) ions. The carbonyl group is oxidized to a carboxyl group. Glucose and other sugars capable of reducing ferric or cupric ion are called reducing sugars. This property is the basis of **Fehling's** and **Benedict's** tests for reducing sugars.

> **Learner's Notes**

Glycosidic Linkage

The classes of carbohydrates other than the simplest monosaccharides, such as disaccharides, oligosaccharides and polysaccharides are made of two or more units of monosaccharide which are joined through a covalent bond, formed by a simple condensation reaction with the removal of a water molecule (dehydration). The covalent bonds that join monosaccharides in disaccharides and complex carbohydrates are called as glycosidic linkage.

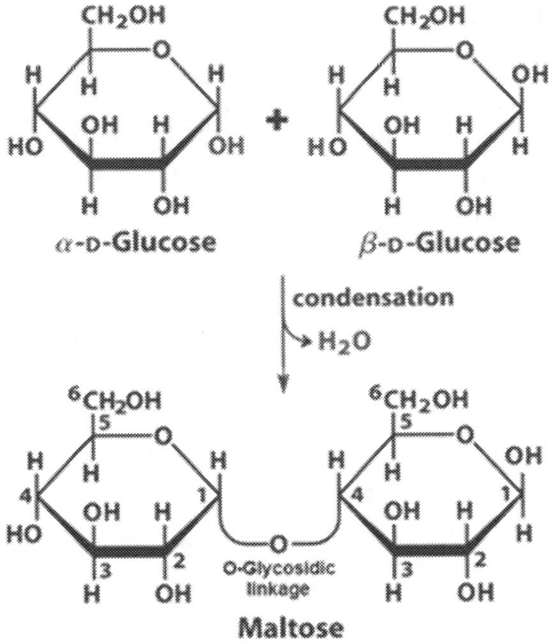

α-ᴅ-Glucose   β-ᴅ-Glucose

condensation

H_2O

Maltose

In disaccharides and polysaccharides, if the anomeric carbon involved in the bond is in α configuration then the resulting glycosidic linkage is called α-glycosidic bond e.g. amylopectin, glycogen. However, If the anomeric carbon is in β configuration, the resulting glycosidic linkage is called β -glycosidic bond e.g. cellulose.

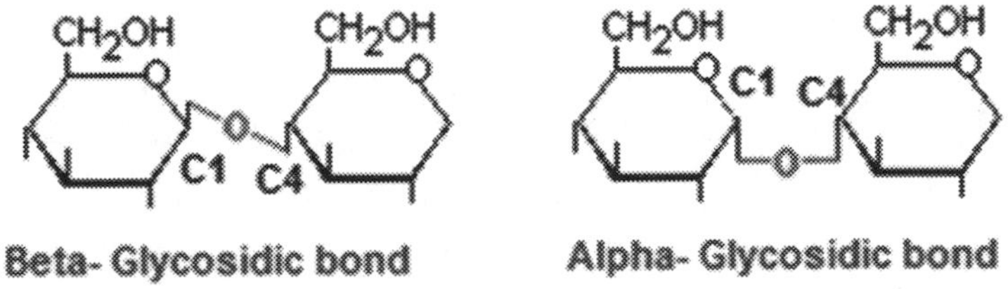

Beta- Glycosidic bond   Alpha- Glycosidic bond

Disaccharides

A disaccharide contains two monosaccharides joined by an *O-glycosidic bond.* The hydroxyl group of one sugar and a hydroxyl group of another sugar or some other compound can react together, splitting out water to form a glycosidic bond.

Monosaccharide + Monosaccaharide → Disaccharide + Water

Lactose, a disaccharide which is a sugar in milk, consists of galactose and glucose bonded with β -1, 4- glycosidic linkage.

Sucrose, a disaccharide commonly called as cane sugar consists of glucose and fructose bonded with α-1, β-2- glycosidic linkage.

Note: *As shown in the structure of sucrose, the anomeric carbons of both monosaccharides are engaged in the formation of glycosidic linkage. In the absence of a free anomeric carbon, sucrose is a non-reducing sugar.*

Maltose, a disaccharide is made up of two glucose molecules bonded with α -1,4-glycosidic linkage.

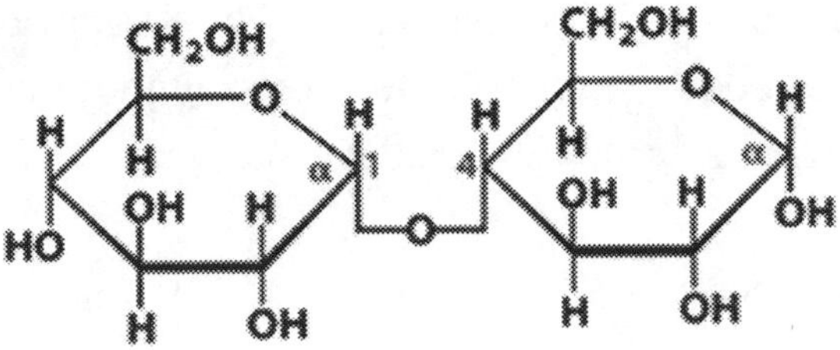

Learner's Notes

Oligosaccharides

Oligosaccharides are complex carbohydrates containing 3-10 monosaccharides linked together by O-glycosidic bond. Oligosaccharides are usually constituents of more complex biomolecules such as lipopolysaccharides and glycoproteins. *Example*: stachyose, raffinose.

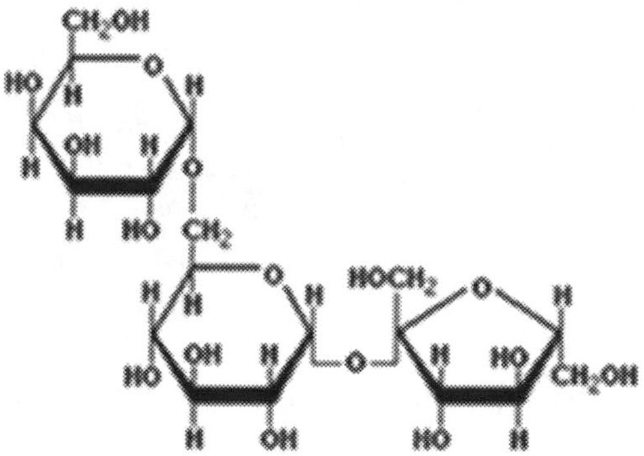

Raffinose

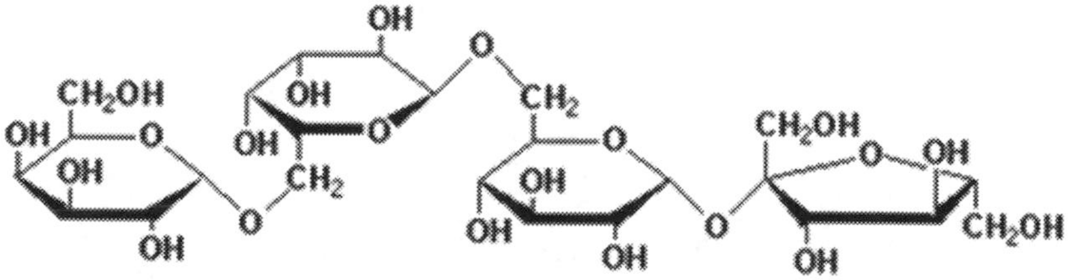

Stachyose

Polysaccharides

These are complex polysaccharides containing tens to thousands of monosaccharide joined by α- or β- glycosidic bonds to form linear chains or branched structures.

Linear polysaccharides: These include complex sugars such as amylose (a form of starch in plant) and cellulose (structural polysaccharide of plant cell wall) which are linear polymers of glucose.

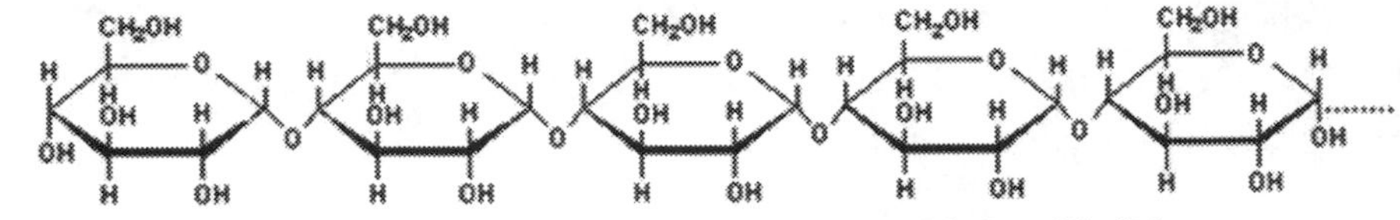

Amylose   α- 1,4 glycosidic linkages

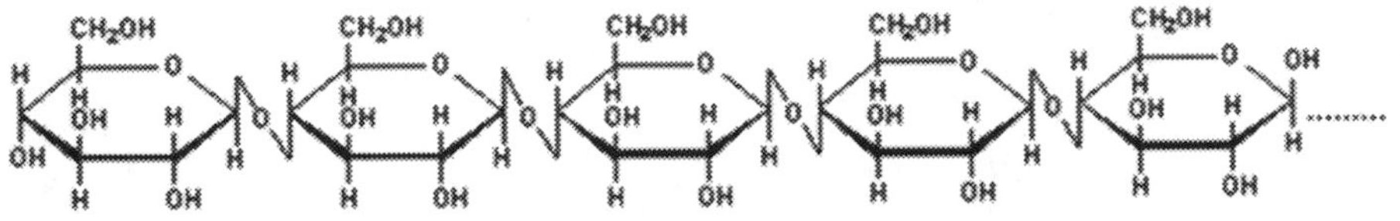

Cellulose   β- 1,4 glycosidic linkages

Branched polysaccharide: These include complex sugars such as amylopectin (a form of starch in plants) and glycogen (the storage form of glucose in animals) which are branched polymers of glucose residues linked through α-1, 4 glycosidic linkages to form linear chains and such chains are branched through α-1, 6 glycosidic bonds.

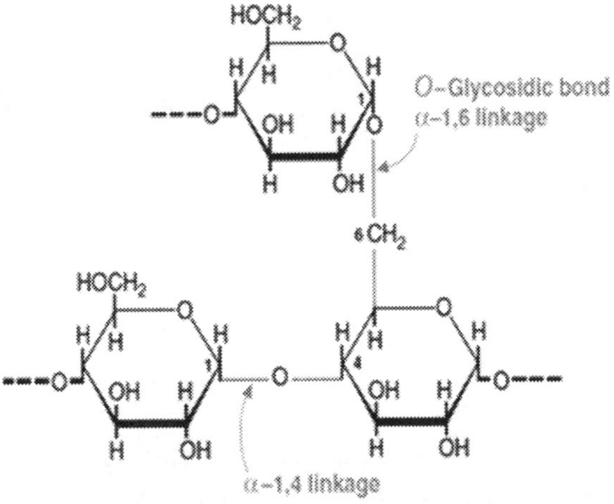

O-Glycosidic bond
α-1,6 linkage

α-1,4 linkage

Learner's Notes

Functions of carbohydrates in living cells

1. Dietary carbohydrates such as **starch (amylose and amylopectin), cane sugar (sucrose), milk sugar (lactose)** and **fruit sugar (maltose)** are primary source of energy for animal cells. In cells such as RBC and brain cells, blood **glucose** is the sole source of energy.

2. Polysaccharide **glycogen** acts as energy reservoir for storing energy that could be used by the cells during the state of fasting.

3. **Cellulose** is a structural carbohydrate that is used by plant cells to make their cell wall.

4. Complex carbohydrate polymers covalently attach to proteins or lipids to form glycoconjugates such as glycoproteins or glycolipids. These molecules serve as recognition signals in cell-cell and cell-matrix interactions. For example; **Neural cell adhesion molecule (NCAM), and antibodies**. Human blood groups A, B, AB and O depends on the **oligosaccharide** part of the glycoprotein on the surface of erythrocytes. The terminal monosaccharide of the glycoprotein at the non-reducing end determines blood group.

5. Some of the secondary functions of carbohydrates include "**protein sparing**" which prevents the breakdown of amino acids for energy needs and "**Fat sparing or preventing ketosis**" which also avoids the breakdown of lipids. It needs to be mentioned that in the absence of carbohydrates, body fulfills its energy demands by breakdown of lipids and proteins.

6. Pentose sugars such as ribose and 2'-deoxyribose are constituents of nucleic acids RNA and DNA, respectively.

Learner's Notes

Proteins

Proteins are polymers (chain) that are made up of smaller units called as amino acid which are linked together by covalent bonds known as peptide linkage.

The general chemical structure of an amino acid consists of a central carbon atom bonded to: a hydrogen atom, a carboxyl group, an amino group and an additional side chain "R" that is unique to each amino acid. The side chain "R" for each amino acid is different and as a result amino acids differ in: shape, size, composition, electrical charge and pH.

Classification of amino acids

1. *On the basis of protein constituents, amino acids are classified as "proteinogenic" and "non-proteinogenic" amino acids.*

Proteinogenic amino acids are those which serve as the building blocks of proteins i.e. proteins are made up of proteinogenic amino acids.

There are 20 proteinogenic amino acids in eukaryotes and these are called as standard amino acids.

$H_2N-CH-C-OH$ $(CH_2)_3$ NH $C=NH$ NH_2 **Arginine**	$H_2N-CH-C-OH$ CH_2 CH_2 $C=O$ NH_2 **Glutamine**	$H_2N-CH-C-OH$ CH_2 **Phenylalanine**	$H_2N-CH-C-OH$ CH_2 OH **Tyrosine**	$H_2N-CH-C-OH$ CH_2 HN **Tryptophan**
$H_2N-CH-C-OH$ $(CH_2)_4$ NH_2 **Lysine**	$H_2N-CH-C-OH$ H **Glycine**	$H_2N-CH-C-OH$ CH_3 **Alanine**	$H_2N-CH-C-OH$ CH_2 N NH **Histidine**	$H_2N-CH-C-OH$ CH_2OH **Serine**
$C-OH$ HN **Proline**	$H_2N-CH-C-OH$ CH_2 CH_2 $C=O$ OH **Glutamate**	$H_2N-CH-C-OH$ CH_2 $C=O$ OH **Aspartate**	$H_2N-CH-C-OH$ $CHOH$ CH_3 **Threonine**	$H_2N-CH-C-OH$ CH_2 SH **Cysteine**
$H_2N-CH-C-OH$ CH_2 CH_2 S CH_3 **Methionine**	$H_2N-CH-C-OH$ CH_2 $CH-CH_3$ CH_3 **Leucine**	$H_2N-CH-C-OH$ CH_2 $C=O$ NH_2 **Asparagine**	$H_2N-CH-C-OH$ $CH-CH_3$ CH_2 CH_3 **Isoleucine**	$H_2N-CH-C-OH$ $CH-CH_3$ CH_3 **Valine**

*Note: There are two standard amino acids in which the side chain-"R" contains sulphur (S). These sulphur containing amino acids are cysteine and methionine.

All these amino acids are represented by either a three letter or one letter code as given below:

AMINO ACID	THREE LETTER CODE	ONE LETTER CODE
Alanine	Ala	A
Arginine	Arg	R
Asparagine	Asn	N
Aspartate	Asp	D
Cystelne	Cys	C

Glutamate	Glu	E
Glutamine	Gln	Q
Glycine	Gly	G
Histidine	His	H
Isoleucine	Ile	I
Leucine	Leu	L
Lysine	Lys	K
Methionine	Met	M
Phenylalanine	Phe	F
Proline	Pro	P
Serine	Ser	S
Threonine	Thr	T
Tryptophan	Trp	W
Tyrosine	Tyr	Y
Valine	Val	V

Non- proteinogenic amino acids are those amino acids which are not the constituents of proteins but serve other independent functions. Example: Ornithine & citrulline are non-proteinogenic amino acids which are involved in the synthesis of urea.

2. *Based on the structure ("R" group/side chain), amino acids can be classified as*:

I). Non-polar, polar and charged amino acids:

Based on the ability of the side chain to interact with polar solvent like water, amino acids may be classified as **non-polar/ hydrophobic/ neutral** (low propensity to be in contact with water) and **polar/ hydrophilic or charged** (easy interaction with water due to formation of hydrogen bonds).

Charged amino acids are further classified as **acidic and basic** amino acids depending on the acidic or basic side chains at neutral pH. Basic amino acids have positively charged side chain at neutral pH due to the presence of nitrogen which gets easily protonated (due to high enough pKa value), gaining a positive charge. Similarly, acidic amino acids have negatively charged side chain at neutral pH due to the presence of carboxylic acid group which easily looses protons (due to low enough pKa value), gaining a negative charge.

Most protein molecules have a hydrophobic core (which is not accessible to solvent) and a polar surface in contact with the environment (which make the proteins soluble in aqueous environment of the cell). While hydrophobic amino acid residues build up the core, polar and charged amino acids preferentially cover the surface of the molecule and are in contact with solvent due to their ability to form hydrogen bonds (by donating or accepting a proton from an electronegative atom).

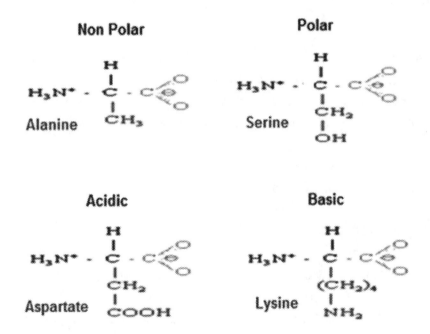

NON-POLAR/ HYDROPHOBIC/NEUTRAL	POLAR/ HYDROPHILIC	CHARGED/ HYDROPHILIC
Alanine Isoleucine Leucine Phenylalanine Valine Proline Glycine	Glutamine Asparagine Serine Threonine Tyrosine Cysteine Methionine Tryptophan	**Acidic** Aspartate Glutamate (Deprotonated form predominant at neutral pH)
		Basic Arginine Lysine Histidine (Protonated form predominate at neutral pH)

II). Aliphatic and aromatic amino acids:

The aliphatic amino acids contain linear or branched hydrocarbon side chains. These are hydrophobic and usually form the core of a protein molecule. The aromatic amino acids have aromatic rings (similar to benzene) in their side chains.

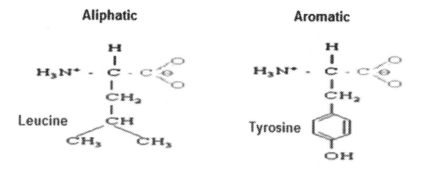

ALIPHATIC	***AROMATIC***
Alanine Valine Leucine Isoleucine	Phenylalanine Tyrosine Tryptophan

3. *Based on nutritional requirement, amino acids are classified as essential or non-essential amino acids:*

Essential amino acids are those which must be present in diet because these cannot be synthesized in our body. The body lacks enzymes that can synthesize these amino acids from any precursor molecules. Non-Essential amino acids need not to be present in the diet because the body can take care of their synthesis when required or when these are deficient in the diet.

ESSENTIAL AMINO ACIDS	NON-ESSENTIAL AMINO ACIDS
Histidine	**Alanine**
Isoleucine	**Arginine***
Leucine	**Asparagine**
Lysine	**Aspartic acid**
Methionine	**Cysteine**
Phenylalanine	**Glutamic acid**
Threonine	**Glutamine**
Tryptophan	**Glycine**
Valine	**Proline**
	Serine
	Tyrosine

* Nutritionally semiessential as synthesized in quantities inadequate to support growth in children

Learner's Notes

Peptide bond formation

Amino acids are linked together by condensation reaction between carboxylic and amino groups from two different amino acids with elimination of water, leading to the formation of an amide linkage called as peptide bond. The product of this reaction is called as a peptide with a *prefix* in accordance to the number of amino acids in the peptide e.g. dipeptide (2 amino acids), tetrapeptide (4 amino acids), octapeptide (8 amino acids) or polypeptide (>50 amino acids).

Peptide bond

Each polypeptide chain starts on the left side by a free amino group of the first amino acid. This end of the polypeptide is termed as amino terminus or N- terminus. Each polypeptide chain ends on the right side by a free carboxyl group of the last amino acid. It is termed as carboxyl terminus or C- terminus.

N- Terminus   **C-Terminus**

Image Courtesy: Dr. William Reusch, Michigan State University

Protein structure

There are four levels of organization in a protein structure (primary, secondary, tertiary and quaternary).

1. Primary structure:

The primary structure of a protein is its unique sequence of amino acids from N-terminus to C-terminus.

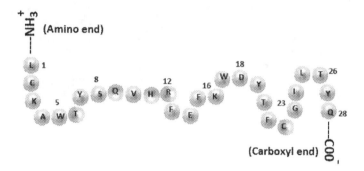

2. Secondary structure:

These results from hydrogen bond formation between hydrogen of –NH group of one peptide bond and the carbonyl oxygen of another peptide bond. According to H-bonding pattern, there are two main forms of secondary structure:

α-helix: It is a spiral structure resulting from the hydrogen bonding between a peptide bond with the fourth one succeeding it i.e.

$1^{st} \rightarrow 4^{th}, 2^{nd} \rightarrow 5^{th}$ and $3^{rd} \rightarrow 6^{th}$ etc.

β-sheets: It is another form of secondary structure in which two or more polypeptides (or segments of the same peptide chain) are linked together by hydrogen bond between H- of NH- of one chain and carbonyl oxygen of the peptide bond of adjacent chain (or segment).

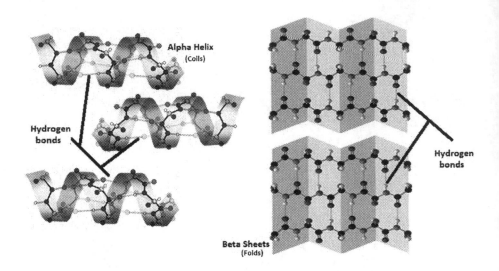

2. Tertiary structure:

This structure of a protein is a three-dimensional organization which is determined by a variety of interactions among the side chains (R-group) of different amino acids or between the R groups and the polypeptide backbone. These interactions include-

Hydrogen bonds among polar side chains
Ionic bonds between R groups of acidic or basic amino acids
Disulfide linkage between cysteine residues of the polypeptide chain
Hydrophobic interactions among non-polar R groups

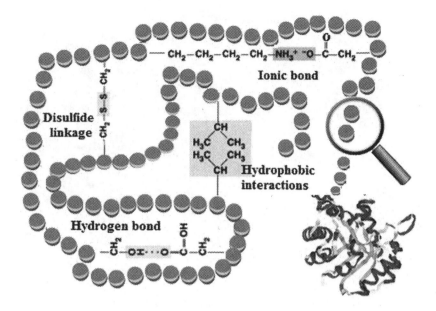

3. Quaternary structure:

It results from the aggregation (combination) of two or more polypeptide subunits held together by non-covalent interaction like H-bonds, ionic or hydrophobic interactions. These structures are present in proteins which are made up of more than one polypeptide subunits. *Examples* of proteins having a quaternary structure are:

> *Insulin*: a protein with two polypeptide chains (dimeric).
> *Collagen:* a fibrous protein of three polypeptide chains (trimeric) that is supercoiled like a rope.
> *Hemoglobin:* a globular protein with four polypeptide chains (tetrameric).

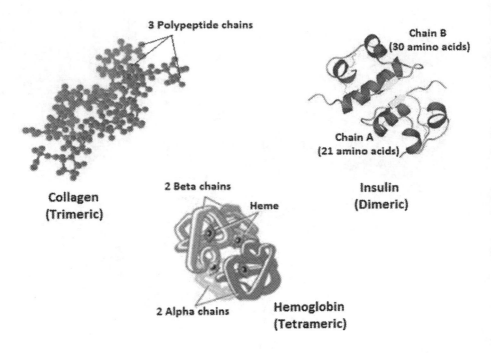

Image Courtesy: Dr. William Reusch, Michigan State University

---

Learner's Notes

Classification of Proteins

Based on their shape, solubility and chemical composition, proteins are classified into the following classes:

(i) Simple proteins: Simple proteins are made entirely of amino acids such that on hydrolysis, they yield only amino acids. According to their solubility, the simple proteins are further divided into *fibrous* or *globular* proteins.

Fibrous Proteins: These proteins also known as scleroproteins are long, thread-like and filamentous polymers of amino acids which serve structural functions in animal cells and tissues. These include water insoluble proteins such as collagen (major protein of connective tissues), elastins (protein of arteries and elastic tissues), and keratins (proteins of hair, wool, and nails).

Globular Proteins: These proteins also known as spheroproteins have the protein polymer folded into a spherical structure. The folding is such that the hydrophilic amino acids form the surface whereas the core is made of hydrophobic amino acids, generally making the globular proteins soluble in water, acids, bases or alcohol. Some examples of globular proteins include albumin (egg protein), globulin (serum), and hemoglobin.

(ii) Conjugated proteins: Conjugated proteins are complex proteins which on hydrolysis yield not only amino acids but also other organic or inorganic components. The non-amino acid constituent of a conjugated protein is called **prosthetic group**. Conjugated proteins are classified on the basis of the chemical nature of their prosthetic groups. These include:

Conjugate proteins	Protein + *Prosthetic group*	Example
Lipoproteins	Protein + *lipid*	Apolipoproteins
Glycoproteins and mucoproteins	Protein + *carbohydrate*	Mucins
Nucleoproteins	Protein + *nucleic acids*	Deoxyribonucleohistones (nucleosomes) Ribonucleoproteins (ribosomes)
Chromoproteins	Protein + *colored pigment*	Hemoglobin (containing heme pigment)

Metalloproteins	Protein + *metals such as iron, copper or zinc*	Hemoglobin, Cytochromes, Ferritin (containing iron); Ceruloplasmin, Plastocyanin (containing copper)
Phosphoproteins	Protein + *phosphoric acid*	Casein

Functions of proteins

1. *Structural functions*: Serves as structural components of animals, e.g. muscle proteins (actin & myosin), eye lens protein (crystallins), bones & skin protein (collagen) and hair & nails protein (keratin).

2. *Functions as hormones:* Some hormones (biochemical messengers) are made up of proteins, e.g. insulin & glucagon (hormones that maintain the blood glucose level).

3. *Functions in immune system:* The Immune response is a series of steps our body takes to mount an attack against foreign invaders (such as bacteria, viruses & parasites). Antibodies are blood proteins that attack and inactivate bacteria and viruses. Major histocompatibility complex proteins (MHC), interleukins and complement proteins play fundamental roles in immune function.

4. *Transport and storage functions:* Protein such as hemoglobin binds and transports the oxygen. In iron metabolism protein ferritin store iron whereas transferrin transport iron.

5. *Biocatalysts (Enzyme):* Enzymes are proteins with globular structure. Enzymes act as biocatalyst to accelerate (increase) the rate (speed) of a biochemical reaction. A restricted region of an enzyme molecule which binds to the substrate is called as *active site*. Enzymes bind to their specific substrates and

convert it into a product at an accelerated rate by lowering the activation energy of the reaction.

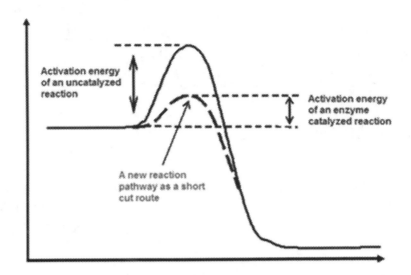

There are six classes of enzymes as classified by the International Union of Biochemistry and Molecular Biology (IUBMB) which has developed a nomenclature for enzymes, the *EC numbers (Enzyme Commission Numbers: 1-6):*

1. *Oxidoreductase:* catalyze redox (oxidation-reduction) reactions by transfer of electrons, e.g. oxidases, reductases, dehydrogenases.

2. *Transferase*: catalyze transfer of a group of atoms from one molecule to another molecule, e.g. kinases, transaminases.

3. *Hydrolase*: catalyze cleavage of covalent bonds by addition of a water molecule, e.g. phosphatases, proteinases, esterases, nucleases, glycosidases.

4. *Lyase*: catalyze the breaking of various chemical bonds by means other than hydrolysis and oxidation, often forming a new double bond, e.g. decarboxylases, synthases.

5. *Isomerase*: catalyze intramolecular rearrangements within a molecule, e.g. epimerases, mutases.

6. *Ligase:* catalyze a reaction in which a C-C, C-S, C-O, or C-N bond is made, e.g. DNA ligase.

These enzymes have been proposed to work through two kinds of mechanism:

The lock-and-key model: The **enzyme** is assumed to be the **lock** and the **substrate** as the **key.** The shape of active site of the enzyme is complementary to the shape of the substrate and the enzyme and substrate are made to fit exactly as lock fits to its key.

The induced-fit model: In this model the shape of active site is not exactly complementary to the substrate. The enzyme active site is a **flexible pocket** whose conformation changes to accommodate the substrate molecule i.e. a change in the shape of an enzyme's active site is induced by the substrate.

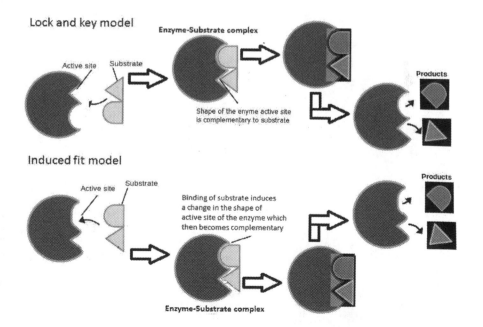

Learner's Notes

Lipids

Lipids are a chemically and biologically diverse group of compounds which are principally defined with a common characteristic feature of insolubility in water. In general, the constitution of lipids are based on molecules derived from fatty acids, four – ring steroid nucleus or isoprene units.

Classification

1. Classification based on function:

Passive lipids: These lipid molecules are engaged in functions which are of quiescent nature. They include storage and structural lipids which make up the bulk of the lipids present in the cell –

Storage lipids: Lipids such as triglycerides (TAG) is stored as fat in adipose cells and serve as the energy-source for organisms and in addition serve as insulating material in the subcutaneous tissues and organs.

Structural lipids: Lipids such as phospholipids and sterols are building blocks of biological membranes.

Active lipids: These lipid molecules constitute minor class in terms of lipid mass of a cell but participate actively in cellular and physiological functions such as:

Signaling molecules: Lipids such as steroid hormones and eicosanoids are messengers in cellular signaling network.

Electron carriers: Ubiquinones such as coenzyme Q form complexes of electron carrier.

Pigments: Carotenoids are lipid derivatives which gives color to vegetables and fruits.

Vitamins: Lipids also serve as micro-nutrients in the form of vitamins such as A, D, E & K.

2. Classification based on chemical composition as proposed by Bloor (1943):

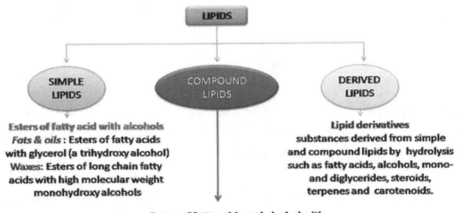

LIPIDS

SIMPLE LIPIDS   **COMPOUND LIPIDS**   **DERIVED LIPIDS**

Esters of fatty acid with alcohols
Fats & oils : Esters of fatty acids with glycerol (a trihydroxy alcohol)
Waxes: Esters of long chain fatty acids with high molecular weight monohydroxy alcohols

Lipid derivatives
substances derived from simple and compound lipids by hydrolysis such as fatty acids, alcohols, mono- and diglycerides, steroids, terpenes and carotenoids.

Esters of fatty acids and alcohol with additional group(s)
Phospholipids: Esters of fatty acids and alcohol with additional group(s) such as phosphoric acid, nitrogen bases and other substituent.
Glycolipids: Esters of fatty acids and alcohol with additional group(s) such as nitrogen and oligosaccharides.

3. Classification based on hydrolysis under basic conditions:

Saponifiable lipids: These lipids contain an ester functional group that can be hydrolyzed under basic conditions. These include triglycerides, phospholipids, glycolipids, sphingolipids and waxes.

Non-saponifiable lipids: These include lipids that are not an ester and cannot be hydrolyzed in basic conditions e.g. cholesterol, bile acids, steroid hormones and eicosanoids.

4. Classification based on the polarity of the lipid molecule

Neutral lipids: These include lipids such as triacylglycerides which do not carry any charge and are thus neutral molecules.

Polar lipids: These include lipids such as phospholipids which attain polarity due to the presence of negatively charged phosphate group. In general, phospholipids have amphipathic character: the 'head' (containing negatively charged phosphate and glycerol) is hydrophilic (attracted to water), while the 'tails' (represented by fatty acids) are hydrophobic (repelled by water).

Basic structural units of lipids

The various classes of lipids have basic carbon skeleton that are contributed by different molecules that include fatty acids (saturated & unsaturated), alcohols (such as glycerol & sphingosine), isoprene units or steroid nucleus.

Fatty acids

Fatty acids are carboxylic acids with hydrocarbon chains ranging from four to thirty six carbon atoms (C4-C36). The most commonly occurring fatty acids have even numbers of carbon atoms in a linear chain of 12-24 carbons. Fatty acids are of two types:

Saturated fatty acids: A long-chain carboxylic acid containing only carbon–carbon single bonds.

Unsaturated fatty acids: A long-chain carboxylic acid containing one or more carbon–carbon double bonds. In naturally occurring fatty acids, the double bonds have *cis* configuration and are unconjugated (i.e. consecutive double bonds are separated by a methylene group).

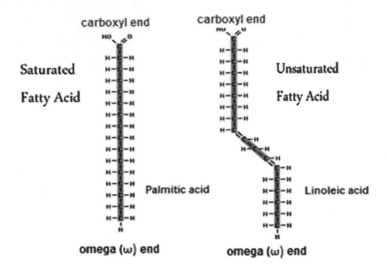

Properties of Saturated Fatty Acids

- ▶ Contain only C–C single bonds
- ▶ Closely packed (hydrocarbon chains in saturated fatty acids are flexible and uniform in shape, allowing them to pack together).
- ▶ Strong attractions between chains
- ▶ High melting points
- ▶ Solid at room temperature
- ▶ Considered bad for health

Properties of Unsaturated Fatty Acids

- ▶ Contain one or more C=C double bonds
- ▶ Non-linear chains do not allow molecules to pack closely (the carbon chains in unsaturated acids have rigid kinks wherever

they contain *cis* double bonds. The kinks make it difficult for such chains to fit next to each other in an orderly fashion).

▸ Few interactions between chains
▸ Low melting points
▸ Liquid at room temperature
▸ Considered good for health (PUFA= polyunsaturated fatty acids also called Omega-3 & Omega -6 fatty acids).

Nomenclature of fatty acids

Fatty acids are designated by the chain length and number of double bonds, separated by a colon. In case of unsaturated fatty acids, the position of double bond (from C1 of carboxyl group) is specified by superscript number following a delta (Δ).

Saturated fatty acid, example; palmitic acid (16:0)

Unsaturated fatty acid, example; linoleic acid ($18:2\Delta^{9,12}$)

Fatty acid -Common Name	Number of carbon	Number of double bonds	Position of double bond (s)	Source
Saturated				
Lauric acid	12	0	-	Coconut oil
Myristic acid	14	0	-	Butter fat
Palmitic acid	16	0	-	Most fats & oils
Stearic acid	18	0	-	Most fats & oils
Arachidic acid	20	0	-	Peanut & corn oil
Unsaturated				
Palmitoleic acid	16	1	Δ^9	Macadamia oil

Oleic acid	18	1	Δ^9	Olive oil
Linoleic acid	18	2	$\Delta^{9,12}$	Vegetable oil
Linolenic acid	18	3	$\Delta^{9,12,15}$	Soybean and Canola oil
Arachidonic acid	20	4	$\Delta^{5,8,11,14}$	Chicken meat & eggs

Essential fatty acids

Linoleic acid (an omega-6 fatty acid) and *α-linolenic acid* (an omega-3 fatty acid) are considered as essential fatty acids since they cannot be synthesized by humans and thus must be supplied through diet. The term omega-3 or -6 indicate the position of first double bond from the omega end of a fatty acid.

Steroid nucleus

The steroid nucleus is a tetracyclic skeleton, consisting of three fused six-membered and one five-membered ring. The four rings are designated A, B, C & D as shown.

Isoprene units

The isoprene unit has the formula $CH_2=C(CH_3)CH=CH_2$. Several units of isoprene can join in a head to head (1-1 link), head to tail (1-4 link) or tail to tail (4-4 link) fashion to form various lipid derivatives.

(1-1 link)    (1-4 link)    (4-4 link)

Alcohols

Polyhydroxy alcohol such as glycerol, long chain monohydroxy alcohol such as triacontanol or amino alcohol such as sphingosine provide carbon skeleton to a number of important lipid molecules.

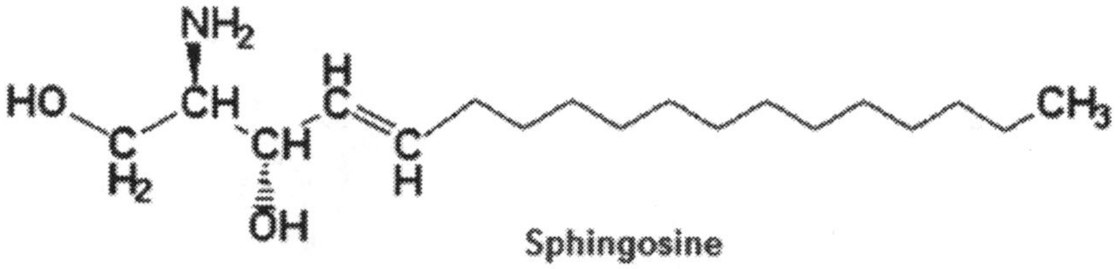

Sphingosine

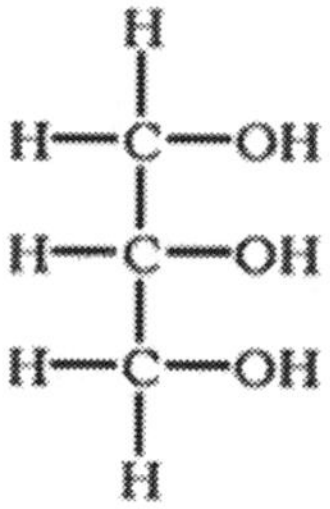

Glycerol

$$HO—CH_2—(CH_2)_{28}—CH_3$$

Triacontanol

Learner's Notes

Important lipids: structure and function

Fats and oils: The triesters of fatty acids with glycerol constitutes the class of lipids known as fats and oils which are primarily the stored form of energy in animals. These triglycerides (TAG) are found in both plants and animals. Triglycerides that are rich in saturated fatty acids are solid at room temperature and are classified as fats which occur predominantly in animals. However, triglycerides that are rich in unsaturated fatty acids are liquid at room temperature and classified as oils, derived mostly from the plant sources.

Simple triglycerides have all the three fatty acids as same such as trilaurin (having three molecules of lauric acid), tristearin (having 3 molecules of stearic acid), triolein (having three molecules of oleic acid). Mixed triglycerides have different fatty acids esterified with glycerol such as distearo-olein (having two molecules of stearic and one of oleic acid), dioleo-palmitin (having two molecule of oleic and one of palmitic acid), stearo-oleo-palmitin (having one molecule each of stearic, oleic, and palmitic acids).

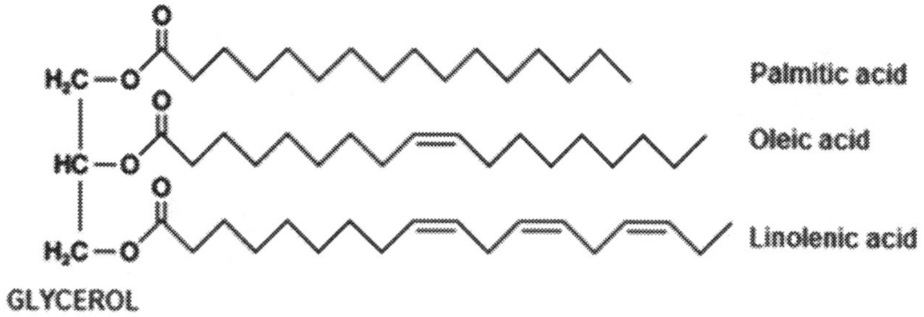

Triacylglycerides

Waxes: These are esters of long chain fatty acids with long chain monohydric alcohols. Example: Triacontanoylpalmitate (a component of beeswax, is an ester of 16C palmitic acid and 30C monohydroxy alcohol triacontanol).

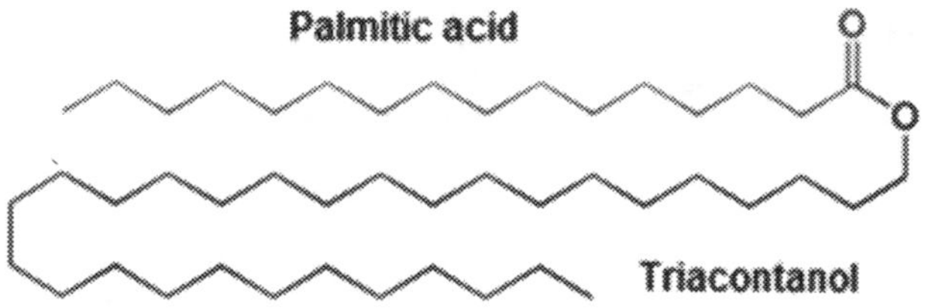

Triacontanoylpalmitate

Sterols: These are structural lipids present in the biological membranes. The characteristic structural feature of sterols is the steroid nucleus. Cholesterol, the major sterol in animal tissues is an amphipathic molecule with a polar head (a hydroxyl group) and a non-polar hydrocarbon body (the steroid nucleus & side chain). Sterols similar to cholesterol in other eukaryotic cells include stigmasterol in plants and ergosterol in fungi. Bile acids (such as taurocholic acid) are polar derivatives of cholesterol that act as emulsifying agents in the digestion of dietary fats. Sterols are also precursors of steroid hormones such as testosterone, estradiol, aldosterone and cortisol.

Cholesterol   Bile acid   Testosterone

Eicosanoids: These are derivatives of 20 carbons polyunsaturated fatty acid called as arachidonic acid. These function as paracrine hormones and are involved in a number of physiological processes such as inflammation, fever, blood pressure, gastric secretions and reproductive functions. There are three classes of eicosanoids which include prostaglandins, thromboxanes and leukotrienes.

Prostaglandin E2   Leukotriene B4   Thromboxane A2

Isoprene derivatives: Lipid molecules like carotenoid pigments such as beta-carotene or lycopene provide color to fruits and vegetables. Lipid derivatives like different forms of vitamin A such as retinoic acid, retinol, retinal serve as essential growth factors and visual elements. The structural constituents of these pigments and vitamin A forms are repeated units of isoprene.

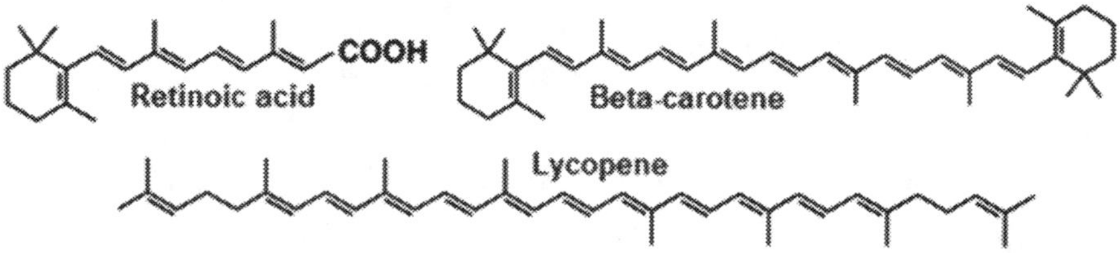

Retinoic acid

Beta-carotene

Lycopene

Phospholipids: Membrane lipids are amphipathic molecules; one end of the molecule is hydrophobic and the other end is hydrophilic. There are two kinds of phospholipids which are the structural units of membrane bilayers. These include glycerophospholipids and sphingophospholipids.

Glycerophospholipids:

These are most abundant lipids in cell membranes which are derivative of phosphatidic acid and are composed of glycerol, two fatty acids (FA), phosphate and a substituent which in most cases is an amino alcohol. The general structure of phospholipids has been shown below where X represents various substituents such as H (phosphatidic acid) or when H is replaced: with -an amino alcohol *ethanolamine* (phosphatidylethanolamine or cephalin), -an amino alcohol *choline* (phosphatidylcholine), -an amino alcohol *serine* (phosphatidylserine),-a *glycerol* (phosphatidylglycerol), -a *phosphatidylglycerol* as substituent (cardiolipin). Phosphatidylcholine is called as lecithin whereas as phosphatidylserine and phosphatidylethanolamine are called as cephalin. Both lecithin and cephalin are abundant in brain and nerve tissues. In all these molecules the two fatty acids form hydrophobic

Mohammad Fahad Ullah

tail whereas phosphate along with the substituent forms a charged polar hydrophilic head.

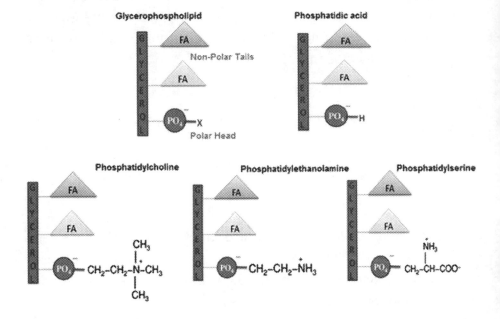

Learner's Notes

Sphingophospholipids:

Sphingolipids are class of lipids similar to glycerolipids but instead of glycerol, these have an amino alcohol called as sphingosine. It has one fatty acid linked by an amide linkage instead of two esterified in glycerolipids. Sphingolipids are classified as sphingophospholipids (having a phosphate with an amino alcohol as in glycerophospholipids), glycosphingolipids (having no phosphate but one or more sugar) and gangliosides (having no phosphates but oligosaccharide chain with one or more residues of N-acetylneuraminic acid). As simplest glycerophospholipid is phosphatidic acid (with H as the substituent), the simplest sphingolipid is ceramide (with H as the substituent).

The general structure of sphingolipids has been shown below where X represents various substituents such as H (ceramide) or when H is replaced: with *phosphocholine* or *phosphoethanolamine* (forming sphingophospholipids called as sphingomyelin that insulates axon of neurons or myelin sheath), *single sugar* such as glucose or galactose (cerebrosides) or *complex oligosaccharides with N-acetylneuraminic acid* also called as *sialic acid* (gangliosides). Gaucher's disease has an accumulation of cerebrosides due to deficiency of beta-glucosidase whereas another disease known as Tay-Sachs disease has an accumulation of gangliosides due to the deficiency of enzyme hexosaminidase.

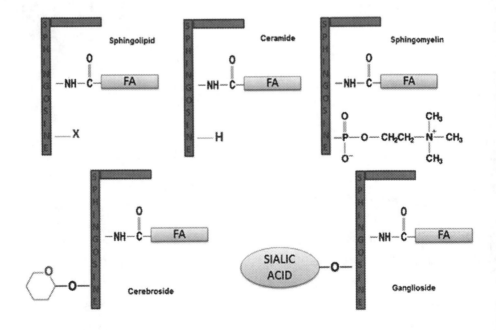

Lipoproteins: Fatty acids (and therefore TAGs) have long hydrocarbon chains which are highly hydrophobic. They can never be transported in our body/blood in "Free" form because of their insolubility in water.

So these are transported across our body in lipoprotein particles in blood.

Lipoproteins are complexes/assemblies of lipids and proteins. Lipoproteins are large, mostly spherical complexes that transport lipids (primarily triglycerides, cholesteryl esters and fat-soluble vitamins) through body fluids (plasma, interstitial fluid and lymph) to and from tissues. Lipoproteins contain a core of hydrophobic lipids (triglycerides and cholesteryl esters) surrounded by hydrophilic lipids (phospholipids, unesterified cholesterol) and proteins that interact with body fluids. So the lipoprotein particles make "insoluble" lipids "soluble" for transportation.

Lipoproteins are classified by density and size, which are inversely related. Density is proportional to low lipid content and high protein content:

1. Chylomicrons (largest size; least density)
2. VLDL, very low density lipoproteins
3. IDL, intermediate-density lipoproteins
4. LDL, low-density lipoproteins
5. HDL, high-density lipoproteins (smallest size; highest density)

Lipoproteins	Triacylglycerol	Cholesterol	Protein
Chylomicrons	>95%	3%	2%
Very Low Density Lipoprotein (VLDL)	70%	20%	10%
Low Density lipoprotein (LDL)	20%	55%	25%
High Density lipoprotein (HDL)	15%	35%	50%

Learner's Notes

Nucleic acids

Structural and functional domains

Nucleotides are the structural and functional units of nucleic acids. There are two kinds of nucleic acids known as deoxyribonucleic acid (DNA) and ribonucleic acid (RNA). RNA is further classified into three functional classes: messenger RNA (mRNA), ribosomal RNA (rRNA) and transfer RNA (tRNA). The amino acid sequence of all the proteins in a cell and the nucleotide sequence of all RNA are specified by the nucleotide sequence of the DNA of the cell. There are a number of cellular functions associated with nucleotides and nucleic acids.

Nucleotides: In addition to their universal function as constituents of nucleic acids, nucleotides also have independent functions that includes; i). Nucleotides such as adenosine tri phosphate (ATP) are energy currency in metabolic transactions in cells, ii). Nucleotides such as cyclic AMP are intracellular signaling molecules acting as second messenger to transmit the signals from hormones and other extracellular stimuli.

Deoxyribonucleic acids: The storage and transmission of biological information are primary known functions of DNA. The segment of a DNA molecule that contains the information required for the synthesis of a functional biological product such as proteins and RNA is called as a gene.

Ribonucleic acids: RNA has diverse function as compared to DNA. There are three types of RNA molecule, each with specific function. Messenger RNAs (mRNA) are the intermediaries that carry genetic information from genes in DNA to ribosome, where the corresponding protein is synthesized. Transfer RNAs (tRNA) are adapter molecules which carry specific amino acids to the ribosome for protein synthesis thereby translating the message encoded on mRNA into protein. Ribosomal RNAs (rRNA) are the structural components of ribosomes which are ribonucleoprotein complexes that carry out the synthesis of proteins.

Nucleotides

DNA and RNA are polymers which are made up of monomers called as nucleotide. In case of RNA, the nucleotide is specifically a ribonucleotide (containing ribose sugar) whereas in DNA these are deoxyribonucleotide (containing deoxyribose sugar).

Components of nucleotide

Nucleotides in both DNA and RNA have three characteristic components: (1) a pentose, (2) a nitrogenous base and (3) a phosphate. The molecule without the phosphate group (containing only pentose and nitrogenous base) is called a nucleoside.

Pentose sugar: The aldopentose sugar ribose is the component of nucleotides. In RNA β-D-ribofuranose forms the pentose skeleton whereas in DNA, it is the derivative of ribose called as 2'-deoxy-β-D-ribofuranose.

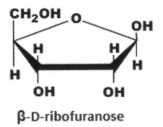
β-D-ribofuranose

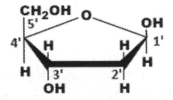
2'-deoxy-β-D-ribofuranose

Nitrogenous base: The nitrogenous bases present in nucleotides are derivatives of two heterocyclic parent compounds, purine (two ring structure) and pyrimidine (one ring structure). Both DNA and RNA contain two kinds of purine bases; adenine (A) and guanine (G). In both DNA and RNA one of the pyrimidines is cytosine (C). However, the second pyrimidine is different in the two nucleic acids with DNA containing thymine (T) and RNA containing uracil (U).

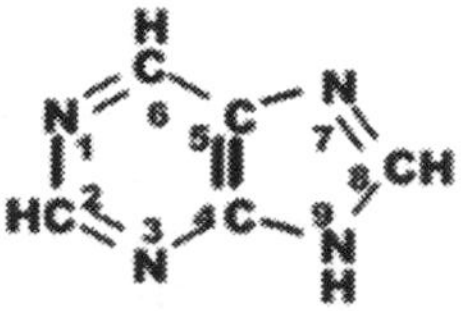

Purine

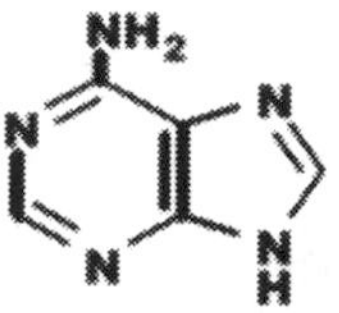

Adenine (A)

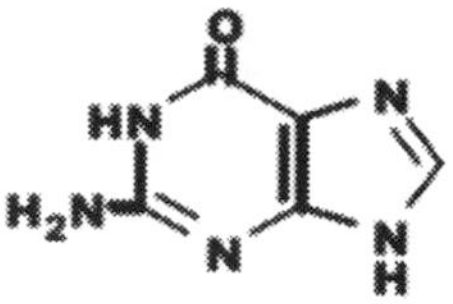

Guanine (G)

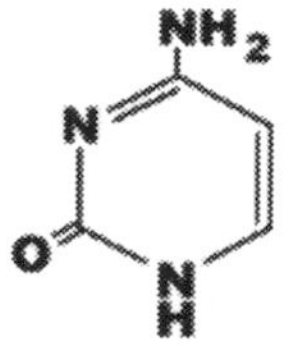

Pyrimidine

Cytosine (C)

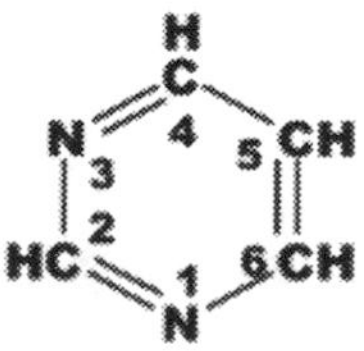

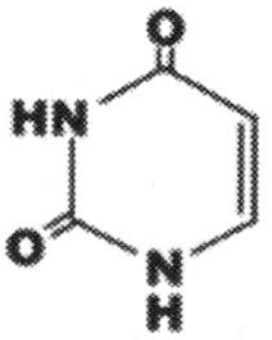

Uracil (U) - RNA

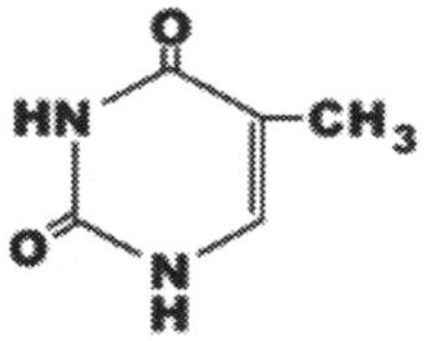

Thymine (T) - DNA

The bases of a nucleotide are linked covalently (at N-1 of pyrimidines and N-9 of purines) in an *N-β*-glycosyl bond to the 1' carbon of pentose while the phosphate is esterified to the 5'carbon. Since both the pentose and nitrogenous base are heterocylic structures, the carbon numbers of pentose are given a prime (') designation as shown below, to distinguish them from the numbered atoms of the nitrogenous bases.

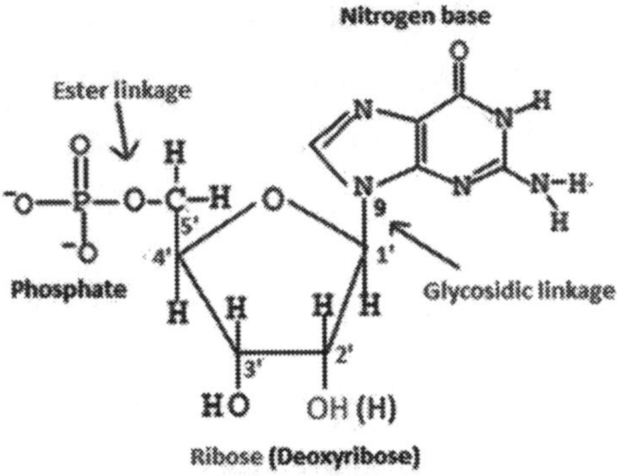

Structure of a nucleotide

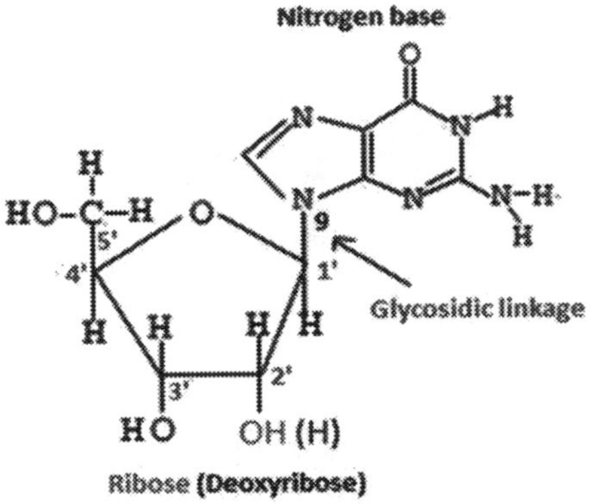

Structure of a nucleoside

Nomenclature: A Nucleoside is named by changing the nitrogen base ending to: **-osine** for purines and **-idine** for pyrimidines. A nucleotide is named using the name of the nucleoside followed by **5'-monophosphate**.

Base	Nucleosides	Nucleotides
RNA		
Adenine (A)	Adenosine (A)	Adenosine 5'-monophosphate (AMP)
Guanine (G)	Guanosine (G)	Guanosine 5'-monophosphate (GMP)
Cytosine (C)	Cytidine (C)	Cytidine 5'-monophosphate (CMP)
Uracil (U)	Uridine (U)	Uridine 5'-monophosphate (UMP)

DNA

Adenine (A)	Deoxyadenosine (A)	Deoxyadenosine 5'-monophosphate (dAMP)
Guanine (G)	Deoxyguanosine (G)	Deoxyguanosine 5'-monophosphate (dGMP)
Cytosine (C)	Deoxycytidine (C)	Deoxycytidine 5'-monophosphate (dCMP)
Thymine (T)	Deoxythymidine (T)	Deoxythymidine 5'-monophosphate (dTMP)

Learner's Notes

Adenosine 5'-monophosphate
Deoxyadenosine 5'-monophosphate

Guanosine 5'-monophosphate
Deoxyguanosine 5'-monophosphate

Cytidine 5'-monophosphate
Deoxycytidine 5'-monophosphate

Deoxythymidine 5'-monophosphate

Uridine 5'-monophosphate

Structure of deoxyribonucleic acid (DNA)

Watson and Crick in 1953 proposed the three-dimensional structure of DNA inspired from the X-ray diffraction images of Franklin and Wilkins. They concluded that DNA is composed of two polynucleotide strands wound around each other in a spiral fashion to form a double helix, with bases on the inside and the sugar-phosphate backbone on the outside.

Each polynucleotide strand is formed by different nucleotides covalently joined to form a long polymer chain by covalent bonding between the 5' phosphate of one nucleotide with the 3' hydroxyl of the sugar of the other nucleotide. Thus, each phosphate-hydroxyl bond is an ester bond and such a bond between two deoxynucleotide is a 3'5' phosphodiester linkage (recall that phosphate group is already bonded to 5' hydroxyl of pentose with an ester linkage and hence diester). Therefore, all of the 5' phosphates and 3' hydroxyls in a polydeoxynucleotide DNA strand are engaged in the phosphodiester linkage except the first deoxynucleotide which has its 5' phosphate free and the last deoxynucleotide which has its 3' hydroxyl free. These free ends provide a DNA strand a polarity; a 5'end (free phosphate) and a 3'end (free hydroxyl). The sequence of nucleotides in a DNA strand is read from the free 5'-end to free 3' end using the letters of the bases.

In the double helix, the two DNA strands are organized in an antiparallel orientation (i.e. the two strands run in an opposite direction, one strand is oriented 5'→3' and the other oriented 3'→5'). The two antiparallel strands are joined by pairing of the bases by hydrogen bonds between the opposite strands. In base pairing, adenine pairs with thymine with two hydrogen bonds (A=T) and guanine pairs with cytosine with three hydrogen bonds (G≡C). This is called complementary base pairing where a large two-ringed purine is paired with a smaller single-ringed pyrimidines to form the most stable conformation. In a DNA molecule the content of purines is always equal to the content of pyrimidines (Chargaff's rule).

The common form of DNA structure is known as **B form**. In this structure of a right handed double helix molecule, the helix makes a turn at every 3.4 nm and has a diameter of 2 nm. The distance between the two neighboring base pairs is 0.34 nm, thereby accommodating 10 base pairs per turn of the helix. The two intertwined strands make two grooves of different width, known as a major groove and a minor groove which facilitate binding of proteins to the DNA molecule.

Under conditions of dehydration such as in the presence of high salt concentrations or alcohol, the DNA attains another form similar to B form (right handed double helix) but with altered dimensions. **A- DNA** is shorter but fatter than B DNA. The helix in A DNA turns at 2.6 nm and the helical diameter is 2.3 nm. The number of base pairs per turn is 11.

Z-DNA is a form of DNA that has a different structure from the more common B-DNA form. It is a left-handed double helix, formed when the DNA is in an alternating purine-pyrimidine sequence such as GCGCGC, and since the G and C nucleotides are in different conformations with regard to sugar (anti & syn), the sugar-phosphate backbone has a zigzag pattern. Z-DNA is taller and thinner than B-form as it has a diameter of 1.8 nm with its helix turning at every 3.7 nm, accommodating 12 base pairs per turn. Z-DNA is believed to play some role in regulation of gene expression.

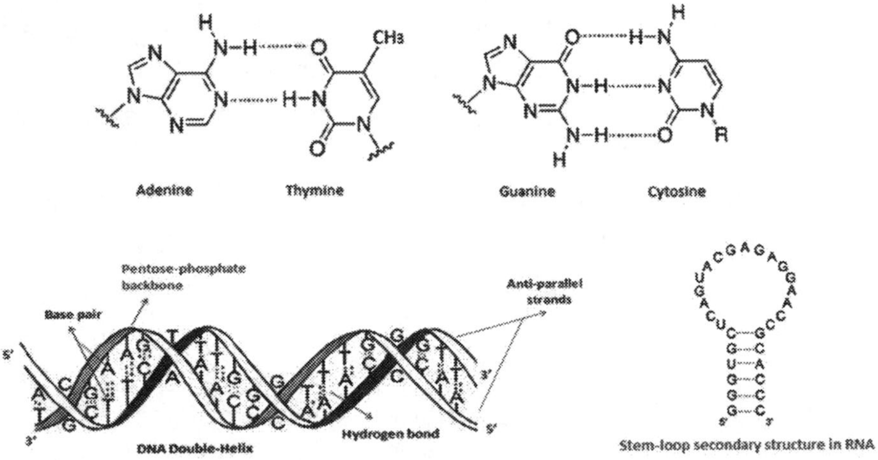

Adenine   Thymine   Guanine   Cytosine

DNA Double-Helix   Stem-loop secondary structure in RNA

Structure of ribonucleic acid (RNA)

Unlike DNA which is double stranded, the ribonucleic acids are single stranded polymer of ribonucleotide covalently joined by phosphodiester linkage. However, in RNA molecules certain regions can form complementary base pairing where the RNA strand loops back to form regions of double strands (as shown above). Transfer RNAs and ribosomal RNAs exhibit substantial secondary structure.

Learner's Notes

Vitamins

Vitamins are organic compounds (small biomolecules) required in trace amounts (microgram to milligram quantities per day) in the diet for health, growth, and reproduction. In humans, almost all vitamins are essential elements, the deficiency of which may lead to disorders. Therefore, these vitamins must be supplied through dietary sources. However, vitamins D and K are considered not as essential since the body can synthesize vitamin D from cholesterol by the action of UV light from the sun whereas vitamin K can be obtained from the microbial normal flora of the intestine.

Classification

The vitamins are classified on the basis of their solubility into two major classes. The *fat-soluble* group (A, D, E, and K) which are soluble in organic solvents and *water-soluble* group including B-complex group vitamins and vitamin C. The fat-soluble vitamins are absorbed, transported, and stored in the body tissues for longer periods of time whereas water-soluble vitamins have a lesser tendency to be retained for long periods of time in the body and show a greater loss by way of urinary excretion.

Water soluble vitamins

Vitamin B complex is a group of water soluble vitamins that include:

a) Thiamine – B1   e) Cyanocobalamin – B12
b) Riboflavin – B2   f) Folic acid
c) Niacin -- B3   g) Pantothenic acid
d) Pyridoxine – B6   h) Biotin

Vitamins of B complex series act as constituents of coenzymes that are necessary for the proper activity of enzymes. Coenzymes derived from these vitamins are small organic molecules that are linked to some enzymes and their presence is essential to the activity of these enzymes. Coenzymes bind to the active site of the enzyme and participate in catalysis without being transformed (changed) at the end of the reaction. Coenzymes often function as intermediate carriers of electrons, specific atoms or functional groups that are transferred in the overall reaction.

Thiamine, also known as vitamin B_1 is a sulfur containing vitamin. It serves as a precursor of the coenzyme thiamine pyrophosphate (TPP) which is required in two general types of reactions: (1) the oxidative decarboxylation of 2-oxo acids catalyzed by dehydrogenase complexes, and (2) the formation of α-ketols (ketoses) as catalyzed by transketolase and as the thiamine triphosphate (TTP) within the nervous system. The deficiency causes beri-beri, a disorder characterized by tenderness of calf muscles, vomiting, tremors, convulsions and loss of appetite.

Riboflavin, also known as vitamin B_2, is an essential component of flavin-adenine dinucleotide (FAD) and flavin mononucleotide (FMN)—coenzymes that are involved in many redox reactions. Flavoproteins catalyze dehydrogenation reactions, hydroxylations, oxidative decarboxylations, deoxygenations, and reductions of oxygen to hydrogen peroxide. Since flavin coenzymes are widely distributed in intermediary metabolism, the consequences of deficiency may be widespread. The deficiency syndrome is characterized by (1) sore throat; (2) hyperemia; (3) edema of the pharyngeal and oral mucous membranes; (4) cheilosis; (5) angular stomatitis; (6) glossitis (magenta tongue); (7) seborrheic dermatitis; and (8) normochromic, normocytic anemia associated with pure red blood cell aplasia of the bone marrow.

Niacin, also known as *vitamin B$_3$*, is converted to the ubiquitous redox coenzymes nicotinamide-adenine dinucleotide (NAD+) and nicotinamide-adenine dinucleotide phosphate (NADP+). The coenzymes NAD+ and NADP+ in which nicotinamide acts as an electron acceptor or a hydrogen donor participate in a large number of redox reactions. Pellagra is the classic deficiency disease of vitamin B$_3$, characterized by chronic wasting disease associated with dermatitis, dementia and diarrhea.

Pyridoxine (pyridoxol), pyridoxamine, and *pyridoxal* are the three natural forms of *vitamin B$_6$*. They are converted to pyridoxal phosphate (PLP), which is required for synthesis, catabolism, and interconversion of amino acids such as in transamination reactions. Deficiency disorders include epileptiform convulsions, dermatitis with cheilosis and glossitis, hematologic manifestations such as decrease in the number of circulating lymphocytes and possibly normocytic, microcytic, or sideroblastic anemia.

Vitamin B$_{12}$, also known as *cyanocobalamin*, is a water soluble hematopoietic vitamin that is required for the maturation of erythrocytes. The generic term vitamin B$_{12}$ refers to a group of physiologically active substances chemically classified as cobalamins or corrinoids. They are composed of tetrapyrrole rings surrounding central cobalt atoms and nucleotide side chains attached to the cobalt. Pernicious anemia is an autoimmune disease that affects the gastric mucosa and results in gastric atrophy. This leads to the destruction of parietal cells and failure to produce intinisic factor, resulting in vitamin B_{12} malabsorption. The deficiency of vitamin B_{12} leads to magaloblastic anemia and neurological disorder.

Folic acid serves as a carrier of one-carbon groups in many metabolic reactions. It is essential for the biosynthesis of compounds such as choline, serine, glycine, purines, and deoxythymidine monophosphate (dTMP). In these reactions, a carbon unit from serine or glycine is transferred to tetrahydrofolate (THF) to form methylene-THF, which is then used as carbon source for: (1) the synthesis of thymidine (and incorporation into DNA), (2) oxidation to formyl- THF to be used in the synthesis of purines (precursors of RNA and DNA), or (3) reduction to methyl-THF, which is necessary for the methylation of homocysteine to methionine. Much of this methionine is converted to S-adenosylmethionine, a universal donor of methyl groups to DNA, RNA, hormones, neurotransmitters, membrane lipids, and proteins. Megaloblastic anemia, characterized by large, abnormally nucleated erythrocytes in the bone marrow is the major clinical manifestation of folate deficiency, although sensory loss and neuropsychiatric disorders are also observed.

Learner's Notes

Biotin (also known as *vitamin H*) is the prosthetic group for a number of carboxylation reactions involved in intermediary metabolism. Such carboxylation reactions catalyzed by carboxylases involve phosphorylation of bicarbonate by ATP to form carbonyl phosphate, followed by transfer of the carboxyl group to the sterically less hindered nitrogen of the biotin moiety. The resulting N (1)-carboxybiotinyl enzyme can then exchange the carboxylate with a reactive center in a substrate. Biotin deficiency may be observed in cases of prolonged consumption of raw egg whites and the symptoms may include anorexia, nausea, vomiting, glossitis, depression, and a dry scaly dermatitis.

Pantothenic acid is a component of coenzyme A (CoA) that is required for the metabolism of fat, protein, and carbohydrate (the citric acid cycle). Pantothenic acid has two major metabolic roles—the first as part of CoA in catabolic pathways and the other as the prosthetic group of the acyl-carrier protein (ACP) in anabolic pathways. Pantothenic acid deficiency has been associated with the syndrome of "burning feet".

Vitamin C is also called as ascorbic *acid*. Like vitamin B complex group vitamin C is classified as water soluble vitamin. Vitamin C plays an important role in tissue oxidation reactions and serves as an anti-oxidant. It is also required for the formation of collagen which is essential for the formation of blood vessels, connective tissues, cartilage and teeth. Vitamin C also helps in the absorption and storage of iron and it is believed to be required for normal growth, tissue repair, healing of wounds and injuries of bones. Scurvy is the deficiency disorder caused due to the inadequacy of vitamin C which is reflected in swollen, tender, and often bleeding or bruised loci at joints, gums and skins.

Fat soluble vitamins

Vitamin A (*retinol*) and its different forms functions as signaling molecules and it is an essential constituent of visual pigment rhodopsin of the vertebrate eye. Retinal combines with the protein opsin to from rhodopsin, which regulates visual transduction by the rod cells of retina to adjust the vision in different conditions of light and darkness. Another derivative of retinol is retinoic acid which serves as hormone that regulates gene expression in the development of epithelial tissues including skin. Deficiency of vitamin A leads to visual impairment with conditions such as xeropthalmia and night blindness.

Vitamin D$_3$ (*cholecalciferol*) is synthesized in the skin from 7-dehydrocholesterol by a photochemical reaction induced by UV light. Vitamin D3 is then metabolized by liver and kidney to its active derivative 1,25-dihydroxycholecalciferol which regulates the absorption of calcium from the intestines and the levels of calcium in kidney and bones. Deficiency of vitamin D leads to defective bone formation in a disorder known as rickets.

Vitamin E (*tocopherol*) contains a substituted aromatic ring and a long isoprenoid side chain. These act as biological antioxidants against reactive oxygen species (ROS) and other free radicals, thereby protecting unsaturated fatty acids from oxidation and preventing membrane lipids from oxidative damage, which can cause cell fragility. Deficiency of vitamin results in fragile erythrocytes.

Vitamin K is called as coagulation vitamin as it promotes synthesis of blood clotting proteins and therefore it is essentially required for clotting of blood. Vitamin K$_1$ or phylloquinone is found in green leafy vegetables whereas vitamin K$_2$ or menaquinone is formed by bacteria residing in the vertebrate intestine. Deficiency of vitamin K leads to hemorrhagic diseases due to the impairment of coagulation mechanisms.

Retinol

Phylloquinone

Tocopherol

Cholecalciferol

Learner's Notes

Metabolites

The living organisms are supported by hundreds and thousands of chemical transformations from one form of organic molecule to another form(s) in a series of enzyme catalyzed reactions, collectively called as metabolism. Metabolism is essential to life and thus the elements of metabolism known as metabolites are also said to be biomolecules.

In general, every metabolic pathway is initiated by a parent molecule or a set of elements, and in the course of several enzyme catalyzed biochemical reactions the parent molecule is transformed into one or more end product(s). All the molecules that are formed in between the parent molecule and the end products are called as metabolic intermediates. These molecules including the parent molecule, the end product and the intermediates which are part of metabolism are said to be metabolites.

The metabolism of biomolecules such as sugar, amino acids, nucleic acids, vitamins, lipids or proteins are required for growth, development and reproduction and hence these are called as *primary metabolites*. The derivatives of primary metabolites which are not involved directly in metabolic processes for growth, development and reproduction but instead have some other roles are said to be *secondary metabolites*. Secondary metabolites may include antibiotics, gums or flavonoids which participate in defense mechanisms or pigments for coloration. *Example:* Glucose which is engaged in energy metabolism is a primary metabolite where as resveratrol, a phytoalexin (defense chemical) in plants is a secondary metabolite.

Mohammad Fahad Ullah

Glucose as primary metabolite

Resveratrol, a stilbenoid as secondary metabolite

Learner's Notes

Essential Readings

Lehninger Principles of Biochemistry 6th Edition by David L. Nelson (*University of Wisconsin-Madison*), Michael M. Cox (*University of Wisconsin-Madison*), Macmillan Publishers.

Tietz Textbook of Clinical Chemistry and Molecular Diagnostics 6th Edition by Carl A. Burtis and David E. Bruns, Elsevier Publishers.

Harpers Illustrated Biochemistry 30th Edition by David A. Bender, Kathleen M. Botham, P. Anthony Weil, Peter J. Kennelly, and Victor W. Rodwell, McGraw Hill Publishers

Marks' Basic Medical Biochemistry 4th Edition by Michael A. Lieberman and Allan Marks, Lippincott Williams & Wilkins Publishers.

About the book

The current edition of this book is intended towards imparting a basic understanding of biomolecules to the students of higher secondary schools and undergraduate programs of health science specialties. An attempt has been made to present the existing knowledge on biomolecules in a lucid language so as to be productive in terms of ease of learning. It is important to mention that the title covers only major categories of biomolecules which are considered to be of significant value in life processes. In short, the book is a compilation of notes for an instant review on the title "BIOMOLECULE".

Printed in the United States
By Bookmasters